AMAZING ANIMALS

SCORPIONS

BY CHRISTOPHER BAHN

I0796439

CREATIVE EDUCATION • CREATIVE PAPERBACKS

Published by Creative Education and Creative Paperbacks
P.O. Box 227, Mankato, Minnesota 56002
Creative Education and Creative Paperbacks
are imprints of The Creative Company
www.thecreativecompany.us

Design by The Design Lab
Art direction by Graham Morgan
Edited by Jill Kalz

Images by Alamy Stock Photo/Ivan Kuzmin, 20; flickr, Biodiversity Heritage Library, 8; Getty Images/Arul Judelin / 500px, 5, Henrik Sorensen, 9, Paul Starosta, 13, 14, 21, PetlinDmitry, cover, 1; Shutterstock/APChanel, 16, Pike-28, 22–23; Unsplash/Erik Aquino, 2, Leon Pauleikhoff, 17, 18; Wikimedia Commons/Pearson Scott Foresman, 7, Toni Wöhrl, 6, Minozig, 10

Library of Congress Cataloging-in-Publication Data
Names: Bahn, Christopher (Children's story writer), author.
Title: Scorpions / by Christopher Bahn.
Description: Mankato, Minnesota : Creative Education and Creative Paperbacks, [2025] | Series: Amazing animals | Includes bibliographical references and index. | Audience: Ages 6–9 | Audience: Grades 2–3 | Summary: "Discover the stinger-tailed scorpion! Explore the arachnid's anatomy, diet, habitat, and life cycle. Captions, on-page definitions, an ancient Greek animal myth, additional resources, and an index support elementary-aged kids"—Provided by publisher.
Identifiers: LCCN 2024011041 (print) | LCCN 2024011042 (ebook) | ISBN 9798889892472 (library binding) | ISBN 9781682776131 (paperback) | ISBN 9798889893585 (ebook)
Subjects: LCSH: Scorpions—Juvenile literature. | Scorpions—Behavior—Juvenile literature.
Classification: LCC QL458.7 .B34 2025 (print) | LCC QL458.7 (ebook) | DDC 595.4/6—dc23/eng/20240415
LC record available at https://lccn.loc.gov/2024011041
LC ebook record available at https://lccn.loc.gov/2024011042

Printed in China

Table of Contents

Scorpions have walked on Earth for 400 million years.

Scorpions are known for their big claws and stinging tail. They belong to a group of animals called **arachnids**. Spiders, ticks, and mites are arachnids, too. There are at least 1,500 kinds of scorpion around the world.

arachnids hard-shelled, eight-legged animals

A scorpion's legs are attached to the front part of its body, called the cephalothorax.

Like all arachnids, scorpions do not have bones. They have an **exoskeleton**. They have eight legs. Their large claws are called pedipalps. A long tail curves over the top of their bodies. It ends in a sharp stinger.

exoskeleton a hard outer shell that supports an animal's body, instead of bones

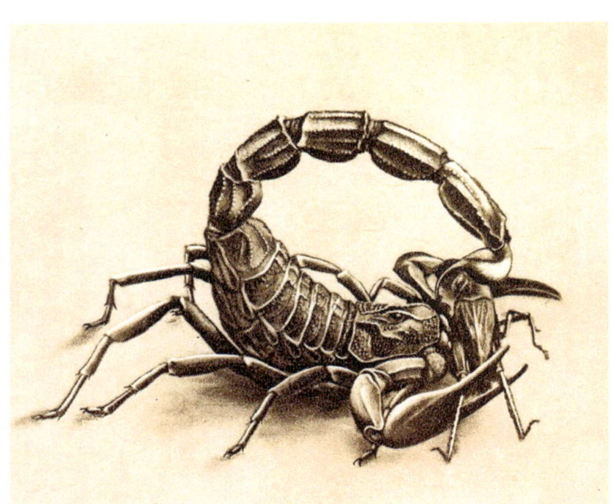

Scorpion tails bend and twist for the perfect strike.

The stinger is the scorpion's main weapon. The animal uses it to hunt and to protect itself. Venom flows through the stinger. Some kinds of scorpion venom are mild. Others, such as bark scorpion venom, can kill a human.

venom a poison made by some animals to harm or kill other animals

Scorpions come in different sizes. Some are shorter than 0.5 inch (1.3 centimeters). Others are 8 inches (20 cm) or more. Smaller scorpions live three to five years in the wild. Larger scorpions may live up to 15 years.

The deathstalker is a medium-sized scorpion and one of the world's most dangerous.

Scorpions are found in warm places, from deserts to rainforests. Some burrow into the ground. Others climb trees. Some scorpions live in caves or on seashores. Cave dwellers are often sightless. They might not even have eyes!

Ground burrows are usually home to just one scorpion.

Scorpions eat insects, spiders, frogs, lizards, and mice—even other scorpions!

Scorpions are predators. They use their speed and stinger to catch food. They do not have teeth. Before eating, they must spray their food with juices from their mouth. The juices turn the food to liquid. Then the scorpions suck it up.

predators animals that kill and eat other animals

Scorpions move the most at night. They try to stay away from direct heat and sunlight. Most scorpions live alone. If they meet another scorpion, they will usually fight it. They may even try to eat it!

Scorpions glow blue or green under a blacklight or ultraviolet light.

Scorpions cannot hear. They don't have ears. Most scorpions have many eyes, but they do not see well. To learn about the world around them, they use special hairs on their bodies. The hairs sense when something passes by.

Fine hairs on a scorpion's body feel movement on the ground and in the air.

Female scorpions give birth 3 to 18 months after mating.

Scorpions "dance"! Before mating, they grab claws and circle around each other. Females give birth to up to 100 babies, called hatchlings. They cling to their mother's back until old enough to molt. Then they live on their own.

molt to shed one's old skin to allow for new growth

A Scorpion Tale

An old Greek story tells of a battle between a hunter named Orion and a giant scorpion. Both died in the fight. To honor the fighters, the gods put pictures of them in the sky. They made the pictures with stars. They placed Orion on one side of the sky and the scorpion on the other. That way, the two would never fight again.

Read More

Adamson, Thomas K. *Scorpion vs. Tarantula*. Minneapolis: Bellwether Media, 2021.

Franchino, Vicky. *Scorpions*. New York: Children's Press, 2015.

Websites

PBS: Deep Look
https://www.pbs.org/video/scorpions-are-predators-with-a-sensitive-side-oaozaf
Watch a video about how scorpions sense the world around them.

Wonderopolis: Are Scorpions Deadly?
https://www.wonderopolis.org/wonder/are-scorpions-deadly
Discover fun scorpion facts and activities.

Note: Every effort has been made to ensure that the websites listed above are suitable for children, that they have educational value, and that they contain no inappropriate material. However, because of the nature of the Internet, it is impossible to guarantee that these sites will remain active indefinitely or that their contents will not be altered.

Index